Virtual Reality and Mechanical Engineering

Innovative Design Methods

Table of Contents

Chapter 1. Introduction

In an increasingly digital world, mechanical engineering, a discipline steeped in tangible operations, is undergoing a paradigm shift through the incorporation of Virtual Reality (VR). This Special Report, "Virtual Reality and Mechanical Engineering: Innovative Design methods," uncovers the burgeoning alliance between these two domains. It provides an in-depth exploration into the cutting-edge design techniques employed, where VR is acting as the conduit between conceptuality and physicality in mechanical design. Presented in a structured, easy-to-understand manner, the report is indispensable for engineering professionals seeking to gain a comprehensive understanding of this innovative intersection, students aiming to stay abreast of modern design methods, and businesses looking to harness the competitive edge VR can offer in mechanical engineering applications. So, if you want to navigate the frontier of this technological integration effectively, this Special Report promises an invaluable foundation.

Chapter 2. Introduction to Virtual Reality and Its Application in Engineering

Virtual reality (VR), a technology that manifests an immersive, interactive, three-dimensional (3D) environment to users, has been rapidly contributing to a wide array of industries - and mechanical engineering is no exception. At its core, VR is about cueing human sensory systems into perceiving a digital environment as if it were real. Leveraging computer graphics, motions sensors, and eye-tracking technology, VR allows users to explore and manipulate virtual space in real-time.

2.1. The Basics of Virtual Reality

The kernel of VR systems is immersive simulation – the creation of a fully digital environment that provides an artificial sensory experience. VR tricks the human brain into perceiving this artificial environment as real. This environment can either replicate a real scenario (such as walking through a planned architectural development) or create an entirely fantastical one (such as exploring an imaginary planet).

VR systems are typically embodied through a head-mounted display (HMD), which tracks the user's head movements and changes the user's view correspondingly. Auxiliary devices, such as haptic gloves, can also be used to provide tactile stimuli and enable interaction with the virtual world.

2.2. Applications of VR across Industries

Though VR earned its initial fame as an entertainment device, primarily in gaming and cinema, its potential applications have multiplied across fields as diverse as psychology, sports, medicine and – notably for this report – engineering. Medical students, for instance, can practice surgeries on virtual patients before operating on real ones. Athletes can run through virtual training scenarios to gain muscle memory.

2.3. The Emergence of VR in Engineering

Engineering, as a sector rooted in innovating practical applications for abstract concepts, stands to benefit immensely from VR. Particularly, mechanical engineering, which deals predominantly with the design, analysis, and manufacture of mechanical systems, finds a complementary partner in VR.

The growing adoption of VR in the engineering industry is driven by several factors. These include an increased recognition of VR's effectiveness as a design tool, through enabling visualisation and interaction with complex 3D models, facilitating collaboration, and allowing prototyping and testing without the need for expensive and time-consuming physical models.

2.4. The Role of VR in Mechanical Engineering

Traditionally, mechanical engineers have relied on two-dimensional (2D) technical drawings to visualise their concepts and share ideas from inception to manufacturing. However, these drawings often become subject to interpretation, leading to misunderstandings and rework costs. By providing engineers with the ability to visualise and work within a 3D environment, VR significantly reduces the possibility of misinterpretation, enhances understanding and increases efficiency.

2.5. Design Visualization and Interaction

The value of VR's immersive nature in mechanical engineering is perhaps most evident in its applications in design visualisation and interaction. Through VR, engineers can virtually interact with the design of a product, machinery or system. They have the ability to 'walk through' the design, examine it from different angles and perspectives, make real-time design adjustments, and even test out the operation of moving parts – all before manufacturing a physical prototype.

2.6. Collaborative VR Environments

Another noteworthy application of VR in engineering involves collaboration. Engineering drawings and CAD models can be uploaded to a shared VR workspace, accessible by engineers working remotely. This virtual meeting room allows multiple engineers to simultaneously view, modify and review a design – in real time and irrespective of their geographical locations.

2.7. VR in Prototyping and Testing

Prototyping is an integral part of the mechanical design process, aimed at testing a design's functionality and identifying any flaws before full-scale manufacturing begins. Conventionally, this process involves physically creating the prototype – a costly and time-consuming exercise. VR, however, allows engineers to create virtual prototypes, thereby accelerating the design process and reducing costs. These prototypes can be tested in various virtual 'real-world' scenarios to identify and rectify any weaknesses or inadequacies.

2.8. Future Prospects of VR in Mechanical Engineering

As VR technology continues to evolve, its future applications in mechanical engineering are both promising and exciting. With advancements in haptic technology, future VR environments could provide mechanical engineers with an even more immersive experience, enabling them to touch and feel the textures and temperatures of different materials. This would greatly enhance their ability to make informed choices about materials during the design process.

Engineers could also make use of machine learning algorithms within VR to predict the performance of their designs under different scenarios, further streamlining the design process and improving the reliability and efficiency of mechanical systems.

In conclusion, the role of VR in mechanical engineering is both impactful and undeniable. It is already revolutionising the way mechanical engineers design, test, and review their creations, forming a crucial bridge between theoretical concepts and practical outcomes. As we continue to explore its potential, VR has the capacity to truly redefine the future of mechanical engineering. This report has begun to delve into that exciting prospect, unravelling the intricate weave of this technologically advanced symbiosis.

Chapter 3. The Basics of Mechanical Engineering Design

Before embarking on an exploration of Virtual Reality (VR) in the realm of Mechanical Engineering Design, it's vital to establish a solid grounding in the fundamentals. At its core, Mechanical Engineering Design is the application of principles from physics, mathematics, and materials science to the analysis, design, manufacture, and maintenance of mechanical systems. This branch of engineering has broad applications, from small individual components to large complex machinery.

3.1. Understanding Mechanical Systems

A mechanical system can be defined as a device or mechanism that transmits power for accomplishing a particular task. Mechanical systems consist of elements such as gears, chains, belts, springs, and other mechanical components. The detailed study of these systems and their behavior forms the backbone of mechanical engineering design.

A common goal when designing such systems is to optimize their overall performance while minimizing costs and ensuring safety. This often involves trade-offs between often competing metrics such as mechanical strength, weight, and durability. However, in the era of digital transformation, new design and simulation tools are transforming how these systems are designed, opening up opportunities for unprecedented innovation and efficiency.

3.2. Principles of Design in Mechanical Engineering

Mechanical Engineering design is dictated by several key principles. These include force analysis, the study of materials and their properties, energy considerations, and manufacturing techniques. Each of these areas possesses its own specialized set of methods and tools.

1. **Force Analysis**: This involves the use of physics to understand how forces interact with objects. Analyzing these interactions can help predict how the system will react under different operating conditions, environmental stressors, or unexpected physical impacts.

2. **Materials Science**: Every mechanical device is built out of materials, each with their own unique properties that can influence the device's performance. Materials science is the study of these properties, and in mechanical engineering design, it informs the selection of materials for each component of a system.

3. **Energy Considerations**: Mechanical systems operate by transforming energy from one form to another (typically mechanical energy into heat or vice versa). Understanding these energy transformations can help identify areas for improvement and optimization in a design.

4. **Manufacturing Techniques**: In mechanical engineering design, the way a product is manufactured can dramatically affect its overall performance. Different manufacturing techniques can introduce variations in the properties of materials, impact the accuracy of the final product, and influence the cost of production.

3.3. The Design Process

The mechanical engineering design process often follows a specific sequence of steps. These can include needs identification, preliminary design, detailed design, and product testing.

1. **Needs Identification**: Before beginning the design process, it is crucial to understand what needs the product is intended to meet. This could include identifying the practical applications, the physical and technical constraints, the target market, and the product's lifespan and maintenance requirements.

2. **Preliminary Design**: This stage involves defining the product's overall layout and shape, its basic functions and components, and materials selection. It's also when any required simulations and experimentation often begin.

3. **Detailed Design**: In the detailed design stage, each component of the product is designed with a high level of detail. This includes decisions regarding materials, tolerances, manufacturing processes, testing specifications, and how the components fit together.

4. **Product Testing**: Once the product has been manufactured, it is rigorously tested to ensure it meets the design specifications and can safely and effectively accomplish its intended function.

These steps are iterative — the results from one stage feed into decisions made at each subsequent stage. This iterative process is intended to maximize the overall quality of the final product and ensure that it meets its intended functions.

In the traditional world of mechanical engineering, the design process, although rigorous and comprehensive, faces limitations, particularly in terms of the time and resources required for iterative testing and enhancement. However, the incorporation of Virtual Reality into the design process promises to redefine these paradigms, offering unparalleled precision, reduced development times, and more lean design and manufacturing methodologies.

Chapter 4. Emerging Intersection: Virtual Reality Meets Mechanical Engineering

In the realm of mechanical engineering, one of the fundamental shifts is transpiring due to the incorporation of Virtual Reality (VR). This novel synthesis presents a dramatic evolution in design mechanisms, optimization processes, as well as testing and validation modalities. Here, we delve deeper into the points of confluence between VR and mechanical engineering, the gravity of these interactions, and the future implications engineered by this creative union.

4.1. Virtual Reality: A Brief Overview

Virtual Reality (VR), an avant-garde technology, creates computer-generated environments or reproduces real-life settings with an unprecedented level of immersive experience. By visually and aurally simulating the surroundings, VR provides an artificial, yet highly sophisticated, sensory experience that can effectively mimic physical presence in the real or imagined worlds. This technology leverages advanced hardware (like headsets and haptic feedback devices) and software systems to render these simulations, pushing the boundaries of user engagement and immersion.

4.2. VR's Application in Mechanical Engineering: Current Status

The application of VR in mechanical engineering, though in its relative infancy, is ushering in a new era of design and development processes. Virtual prototyping - the creation of digital, three-dimensional designs of products - is one of the direct applications already operational in the industry. It brings substantial cost reduction and time efficiency, eliminating the need for multiple physical prototypes. Design errors can be identified and rectified early in the process, thus improving the overall product quality and reducing time-to-market.

Furthermore, VR facilitates virtual assembly and disassembly sequence planning, significantly slashing the time invested in real sequence planning. Engineers can manipulate virtual parts and assess the repercussions of different assembly or disassembly sequences. This capability mitigates risks, improves safety, and provides a deeper understanding of the product even before it physically exists.

4.3. Design Optimization through Virtual Reality

Using VR, engineers can move away from the traditional 2D CAD models and instead work with 3D models within a virtual environment. These spatially interactive models provide a superior understanding of the design and its functionality. This enhanced perspective ensures optimization of the design process right from the initial stages, reducing inefficiencies and identifying potential issues ahead of time.

Simultaneous Multi-user collaboration, another VR hallmark, allows engineers from various disciplines to interact with the design simultaneously, without geographical limitations. They can physically 'walk' through the design, make real-time changes, experiment with different scenarios, all within a shared virtual environment. This collaborative approach accelerates the design process, promoting knowledge sharing and avoiding conflicting design elements.

4.4. Simulation and Validation: Ensuring Safety and Efficiency

Usually, real-world testing of mechanical systems can be expensive, time-consuming, and, at times, perilous. VR addresses these issues by replicating realistic conditions within a virtual environment for safe, affordable, and rapid testing. From testing complex machinery to simulating intricate operational conditions, VR makes these processes far more streamlined and effective.

This technology isn't limited to products and systems; it's also transforming the field of ergonomics. VR can simulate human interaction with the machinery during the design phase, it's possible to evaluate human factors such as reachability, usability, and operator comfort without risking harm to any individuals.

4.5. VR in Education and Training

Virtual reality brings a transformative edge to educational experiences. Mechanical engineering students can explore complex concepts, inspect machinery in intricate detail or conduct 'hands-on' practical work, all within a risk-free, virtual space. For instance, VR can simulate the disassembly and reassembly of an engine, which students can repeat until they master the skill, without the risk of damaging expensive pieces of equipment.

Similarly, VR proves pivotal in training and skill development for existing professionals. For training sessions related to safety procedures, equipment operation, or complex assembly processes, VR-based programs provide an immersive, interactive, and safe training environment.

4.6. Looking Ahead: The Future of VR in Mechanical Engineering

The ongoing synergy between VR and mechanical engineering promises to revolutionize the industry. With the continuous advancements in VR technology: improved graphic realism, lower latency, more accurate motion tracking and stronger haptic feedback; the applications in mechanical engineering will further expand.

Aside from the already-in-play uses in design, simulation, optimization, and testing, future implications are extensive. Whether it be remote operation of heavy machinery, seamless digital twinning, or creating VR-based digital factories, the prospects are limitless.

To summarize, the pivotal role played by VR in the innovative design process within mechanical engineering cannot be overstated. It's bridging the gap between conceptualization and realization, making the digital tangible, and serving as the catalyst for a new age in mechanical engineering. Virtual reality is rapidly becoming a reality within the discipline - a transformative reality that reshapes the landscape towards efficiency, safety, and disruption.

Chapter 5. VR-Driven Conceptual Design in Mechanical Engineering

Virtual Reality (VR) has been making waves across industries, and none more so than in the mechanical engineering sphere. As product design grows to become more complex and multi-faceted, engineers are increasingly resorting to digital techniques that offer precision, flexibility, and interactivity. In step with a new generation of design methods, VR-based conceptual design forms a crucial part of this revolution.

5.1. VR and Mechanical Engineering: A Prime Intersection

Virtual Reality couples digital technology advancements with mechanical design processes, offering a vibrant, immersive experience. Pioneered by technologies like head-mounted displays (HMDs), haptic feedback devices, and stylus-based systems, VR can rapidly and realistically simulate a wide array of design scenarios, enabling mechanical engineers to work in a virtual yet tangible design environment. The benefit of this duality is multi-fold: it allows engineers to better conceive, visualize, modify, and test their designs virtually before they become tangible products in the real world.

5.2. The Power of Visualization and Interactive Design

The first step in designing any product is visualization. By providing a 3D, spatial, and interactive representation of an idea, VR brings visualization to another level. Engineers can develop and modify designs within a virtual realm, considering ergonomics, aesthetics, and feasibility from the get-go. VR also allows 'walk around' and 'inside' the product, offering invaluable perspectives that were, until now, hard to attain. This advancement in visualization capabilities leaps beyond static images and 2D drawings, fueling a new wave of product development efficiencies.

5.3. VR-Driven Prototyping: An Evolution

Prototyping is a critical stage in the mechanical design process. Traditionally, it involves significant costs, time, and resource consumption. However, VR introduces a transformative way of prototyping that circumvents traditional pitfalls. By creating a 'Virtual Prototype' before the physical one, engineers can identify and rectify design flaws early, leading to enhanced product quality, reduced iterations, and lower manufacturing costs. The fidelity of these virtual prototypes is so high that engineers can even study the interactions between different subsystems, further strengthening the overall design.

5.4. User Testing Modernized

VR equips engineers with an extremely powerful tool for user testing. Traditional user testing methods are often unable to accurately predict usability issues due to the challenge of effectively mimicking real-world environments and situations. However, VR offers an immersive environment that closely replicates reality, permitting accurate understanding of how the product will be used, interacted with, and perceived by the user. This ability to gain early and insightful user feedback is crucial to the design process, facilitating more targeted modifications and resulting in higher user satisfaction with the final product.

5.5. Collaborative Design Facilitated

The collaborative potential of VR is another standout feature. VR platforms can support multiple users concurrently, irrespective of their geographic locations. Millennials, the dominant faction in the current workforce, are more inclined toward collaborative work environments. VR interfaces with collaborative features perfectly meet this evolving work culture, fostering creativity, promoting thorough problem-solving, and expediting the design process.

5.6. Virtual Training Real-Life Applications

The final piece of the VR integration into mechanical engineering lies in training. With VR, training is no longer confined to physical confines or theoretical knowledge. VR facilitates a hands-on, real-time experience wherein trainees can interact with equipment or machinery without genuine risk. The ability to perform operations in a safe, virtual environment before actual application significantly enhances learning and understanding, making VR an essential asset for mechanical engineering education.

In sum, the intersection of VR with mechanical engineering is an evolutionary leap in technological innovation and design methodology. By encompassing the entire design process - from conceptualization to post-production testing - VR transforms the way mechanical engineers work. Design becomes more realistic, collaborative, and accurate, leading to high-quality, cost-effective outcomes. For individuals and firms working within this space, staying abreast of this developing landscape is no longer a choice, but a necessity. Adapt or get left behind; VR-driven design is the future.

Chapter 6. Immersive Simulation and Testing with Virtual Reality

Virtual reality (VR) technologies have not only revolutionized the way mechanical engineers design products, but they have also significantly advanced the procedures for simulating and testing device performance and reliability. By facilitating a completely immersive environment, VR enables engineers to interact with their designs in a three-dimensional space, driving innovation by enabling product testing under realistic scenarios before materializing the concept.

6.1. Role of Virtual Reality in Simulation and Testing

Mechanical engineering relies heavily on complex simulations and meticulous testing to optimize product design and performance. Virtual reality enhances this process by providing an immersive, interactive environment where engineers can assess their designs under diverse conditions. This VR-enabled setting allows for the operation of prototypes in a fully synthetic but realistic atmosphere, reducing expensive, time-consuming physical prototype iterations.

Virtual Reality, by bridging the gap between digital design and actual production, extends the possibilities of simulation beyond CAD models on a screen. With VR, engineers can 'touch,' 'feel,' and 'operate' their designs. For example, an engineer can explore an engine's internal parts from unique perspectives, enabling them to understand how modifications in design parameters might influence its functionality.

6.2. Iterative Design Advancements

VR advances the iterative model of design - a repetitive method of prototyping, testing, analyzing, and refining a product. With traditional methods, each iteration demands considerable time and resources. However, with VR, the timeline and cost associated with each iteration dramatically decrease. Engineers can test multiple design versions simultaneously in a virtual environment, identifying potential problems early in the developmental process. As a result, design flaws are detected and rectified early, enhancing the product's reliability and efficiency while reducing costs.

These advancements can be seen not only in the exploration of critical design parameters but also in human factors engineering. An immersive VR environment allows designers to evaluate how users might interact with a product, identify user-related issues, and iterate on the design accordingly. For example, assessing vehicle ergonomics becomes more efficient as adjustments to the driver's seat position or dashboard layout can be tested interactively and modified in real-time.

6.3. Advanced Prototyping

VR allows mechanical engineers to create more sophisticated and intricate prototypes. The conventional prototyping process requires physical prototypes, which can be time-consuming, costly, and, in some cases, even impossible due to complexity or safety reasons, such as nuclear reactor designs. However, VR negates these limitations by providing an immersive environment for advanced prototyping.

Engineers can now simulate operational conditions comprehensively, experiencing their designs dynamically. They can manipulate individual components, run simulations, and observe how various elements react. Engineers can even 'walk inside' their designs, enhancing their understanding and perspective of the real-world complications their products may face.

6.4. Virtual Training

Virtual training scenarios offer another dimension to VR's influence in simulation and testing within mechanical engineering. VR training modules can help engineers and operators understand the intricate operation of machinery and systems in a safe, controlled environment. These modules can replicate operational processes intric enough, enabling individuals to understand the effects of operational or control changes on the system entirely.

For instance, VR simulation can facilitate an in-depth understanding of a jet turbine's operation. Trainees can observe the process from within, understanding the systems in action, isolating each component's operations, and enhancing their operational proficiency without actual operation risks.

6.5. Limitations and Future Directions

Despite the enormous potential of VR in mechanical engineering, there are limitations. The accuracy of VR simulations depends on the realism and detail of the virtual environment, demanding substantial computing power and advanced rendering algorithms. Also, physically interacting with a virtual model can still be challenging due to the lack of haptic feedback.

However, ongoing advancements in VR technology, like haptic gloves for tactile feedback and more powerful processing technologies, promise to fill these gaps. Mechanical engineers can look forward to even more immersive, accurate simulations and testing in the coming years.

In closing, VR's incorporation into mechanical engineering for immersive simulations and testing offers immense potential. The ability to manipulate, observe, and test designs in a safe, yet realistically mimicked virtual environment reduces costs, boosts efficiency, and propels innovation. Evidently, VR is set to be a cornerstone technology in future mechanical design processes, pushing the boundaries of what is possible in simulation and testing.

Chapter 7. Real-world Case Studies: Successful Implementation of VR in Mechanical Engineering

Mechanical engineering and VR have been rapidly converging, creating a dynamic ecosystem where traditional boundaries are being pushed, leading to astounding breakthroughs. This chapter aims to delve into real-world case studies that showcase the remarkable success of integrating VR into mechanical engineering projects.

7.1. Volkswagon's Virtual Reality Training

In 2017, Volkswagen started a pilot project converting 30 of their conventional training programs into VR modules. Trainees would don HTC Vive headsets to allow for interactive, real-time training without the risks or logistical nightmares often associated with high-precision instruction. The results of this initiative were striking, with Volkswagen reporting improved comprehension, retention, and accuracy among the trainees, leading to higher productivity and safety. This case highlights the ability of VR to provide a comprehensive and immersive training interface in mechanical engineering, thereby augmenting skills development.

7.2. BMW & Unreal Engine: Immersive Design Exploration

BMW embraced the integration of VR into their designing phase with the help of the Unreal Engine, a highly versatile game engine known for its use in creating hyper-realistic virtual environments. BMW engineers could don a VR headset and virtually "walk" around their designs, interacting with the model in real time, inspecting intricate sections, and conducting analysis that would not be possible with traditional CAD models. This application of VR in the design phase fast tracks prototyping, reducing financial and temporal expenditures while increasing overall efficiency and productivity.

7.3. Shell's VR Safety Training

Shell implemented VR technology as a training tool for their workers in different drilling scenarios. In real life, a minor mistake in this high-risk environment could trigger catastrophic events; however, within the safety of VR, engineers can simulate and interact with a whole host of potential scenarios. Routinely practicing in an immersive virtual reality environment, as opposed to traditional static simulations, enhances the engineers' skills and response times, thus substantially improving workplace safety.

7.4. Daqri Smart Helmet: Augmented Reality

While not strictly VR, Daqri's smart helmet merges the physical and digital worlds similarly, providing a classic case study for AR (Augmented Reality) in Mechanical Engineering. The helmet combines thermal vision, data visualization, camera recording, and virtual instructions overlaid in the real world, making it an essential tool for complex mechanical processes. It symbolizes how VR and AR can merge to create multi-faceted digital tools for mechanical engineering applications, leading to higher productivity and safety standards.

7.5. NASA's Hybrid Reality Lab

NASA's Hybrid Reality Lab employs a blend of VR and AR to simulate the International Space Station's (ISS) environment for astronaut training. With the added tactile elements, users can experience and react to different scenarios that would be otherwise costly, logistically challenging, and risky to recreate physically. The immersive simulation creates a comprehensive training system that prepares astronauts more effectively for extra-terrestrial working conditions.

In conclusion, the above case studies underscore the transformative impact of VR in Mechanical Engineering when used in design, training, safety simulations, and operations. As the technology continues to evolve, diverse industry players such as EMS providers, automotive manufacturers, oil & gas companies, and space agencies are well-poised to leverage VR for their benefit, thus continually reshaping the modern mechanical engineering landscape. Remarkably, the full potential of this powerful medium remains to be realized as more sectors embrace this exciting convergence.

Chapter 8. Training and Education in Mechanical Engineering through VR

Modern technological developments are greatly affecting the dimensions of learning and instruction in every field, with mechanical engineering being no exception. Therefore, it becomes imperative to shed light on how education and training in this discipline are revamped through next-generation tech like Virtual Reality.

8.1. VR - An Integral Pedagogical Tool for Engineering Education

Virtual Reality (VR) is steadily carving its niche as an instructive tool given the unique experiential learning environment it enables. Unlike traditional classroom lectures, VR provides an immersive, interactive, and three-dimensional platform, facilitating an enriched understanding of complex teachnical concepts.

Table 1. Aosciidoc Table

Advantages	Disadvantages
Enhanced Engagement	Cost of Equipment
Hands-on Experimentation	Technical Difficulties
Contextual Learning	Possible Health Impacts

This table shows the advantages and disadvantages of employing VR for engineering education. Casual observers might view the cost and potential health impacts as considerable downsides. However, as VR becomes more accessible and researchers continue to refine its application, these challenges are predicted to diminish in coming years.

8.2. Integrating VR into Mechanical Engineering Curriculum

Several strategies should be adopted to exploit the strengths of VR and integrate it into mechanical engineering curricula for optimum learning outcomes.

- Tailored VR Simulations: These simulators should aim to break down complex concepts into simple, easy-to-understand interactive modules.

- Real-life Experimental Scenarios: Students should be provided with VR versions of practical experiments eliminating the potential hazards of real-world laboratories.

- VR-based Assessments: Quizzes and tests built inside the VR environment can provide more context to the testing process, enhancing the learning outcomes.

- Industry Partnerships: Collaborations with industries for VR co development can lead to more relevant and job-ready skills for students.

8.3. Case Studies of Successful VR Integration

Two universities have stood out recently for their use of VR technology in mechanical engineering: the University of New South Wales (UNSW) in Australia and the University of Sheffield in the UK.

The UNSW has utilized VR technology in teaching the subject of Thermodynamics. By creating a graphically complex power plant inside a VR environment, students can 'move around' the facility and witness processes happening real-time, thus bypassing the safety regulations and cost associated with a real power plant tour.

The University of Sheffield, meanwhile, has adopted VR technology in their 'Diamond Simulator,' a tailored VR-based teaching tool. This simulator assists the students in grasping the underlying principles of a gas turbine, in a controlled and safe environment.

8.4. Preparing for a VR-Driven Future in Mechanical Engineering

While VR usage in education is still in an early stage, institutions need to start preparing for a future where this technology could be ubiquitous.

Here are some steps that can be taken:

- Training Professors: Programs should be undertaken to familiarize educators with VR technology and its capabilities.

- Infrastructure Development: Institutions should invest in VR tech and infrastructure, ensuring students, irrespective of their financial means, can access it.

- Encourage Research: Facilitate studies into the effectiveness, implementation challenges, curriculum development, and long-term impacts of VR in education.

- Monitoring VR Trends: Continually investigating new developments in the VR industry can ensure the educational implementations stay relevant and cutting edge.

Ahead lies a period of transition and adaptation to VR's capabilities and benefits, ultimately leading to a more dynamic, engaging, and experiential learning environment. By harnessing this power, mechanical engineering education stands to witness a significant transformation into a more practical, interactive, and relevant discipline for the next generation of engineers.

Chapter 9. The Future of Mechanical Engineering Design: Predictions and Trends

Mechanical engineering and its traditional design methods have been the backbone supporting the manufacturing industry. But, as we step into a future heralded by innovations like Virtual Reality (VR), the field is experiencing a significant transformation. This chapter attempts to put forth a broader view of the likely trends and predictions that are set to redefine mechanical engineering design in the future.

9.1. The Advent of Virtual Reality

Virtual Reality (VR) has undoubtedly been one of the most transformative technologies brought to the mechanical engineering spectrum. VR's underlying advantage lies in its ability to create a fully immersive 3D environment, which lets engineers visualize, manipulate, and interact with the design elements in ways that were elusive before. The potency of VR allows designers to go beyond the limited display of conventional CAD tools and determine the real-world implications of designs, leading to safer and more efficient products. By 2030, it's projected that VR-based planning, prototyping, and testing will become the standard in mechanical engineering companies.

9.2. Incorporation of AI in Mechanical Design

Artificial Intelligence (AI) is set to make a significant impact on mechanical engineering design. AI can quickly assimilate large amounts of data and use the resulting insights for more accurate design models. Soon, mechanical engineers will be able to harness the capabilities of machine learning (ML) and AI to craft designs that not only meet the specified requirements but also boast of enhanced efficiency and sustainability. Predictive analytics powered by AI will also form the backbone of future design optimization, allowing for development cycles to be faster and more cost-effective.

9.3. Rise of Additive Manufacturing

Additive manufacturing, otherwise known as 3D printing, is becoming a synonym for flexibility and versatility in mechanical design. The technology allows engineers to fabricate complex geometries that would be extremely arduous, if not impossible, through traditional subtractive manufacturing processes. Engineers can leverage this technology to produce prototypes swiftly and make design changes with reduced overheads. With the maturing of metal 3D printing technologies, the manufacturing of end-use parts will gain more prominence in the industry by 2040.

9.4. Integrated Design and Manufacturing

Integration of design and manufacturing is gaining momentum in mechanical engineering. Traditionally, the design phase and manufacturing phase were separated entities - a methodology that often led to a lengthy and costly development cycle. However, technologies such as IoT (Internet of Things) and digital twins are enabling real-time data exchange between design and manufacturing units. It is leading to quicker iterations and superior coordination, resulting in higher productivity and cost-saving. By 2035, 90% of mechanical engineering enterprises are likely to have achieved this integration.

9.5. Enhanced Collaboration Predicted

Technological advancements like cloud computing are fostering possibilities for seamless collaboration within the mechanical design domain. With these advancements, teams can work simultaneously on design iterations irrespective of geographical location. The concept of remote working in mechanical design will become more entrenched as businesses adapt to the changing work environment models.

9.6. Sustainability in Design

The urgency to promote sustainable practices in every sector has started to make its mark in mechanical engineering design as well. Engineers are focusing more on creating designs that have minimal environmental impact without compromising the efficiency of the product. For instance, Life Cycle Assessment (LCA) tools are gaining prominence in assisting the greening of mechanical engineering. By 2050, a major shift towards eco-design is expected within the field, promoting a circular economy.

9.7. Upcoming Regulatory Changes

As new technologies become more prevalent, they will inevitably bring along regulatory changes. Factors such as product safety, environmental impact, and users' privacy will be at the forefront of these changes. Mechanical Engineers and design professionals need to keep an eye on these evolving regulations and make sure their designs abide by them.

To say predictability is a challenging phenomenon would be an understatement. However, the trends highlighted here have already begun to reshape the landscape of mechanical engineering design. While the digital revolution has significantly transformed the field, the future advancements promise even more exciting opportunities. Undoubtedly, the incorporations of technologies like VR, AI, and 3D printing are enabling a more sophisticated, efficient, and greener industry.

It requires an open mind and a learning mindset from all stakeholders to successfully navigate this rapidly changing scene. It's crucial that mechanical engineering professionals, students, and businesses stay updated with these evolving trends to helm their operations efficiently and anticipate the market needs more accurately, thereby gaining the competitive edge they seek.

Chapter 10. Challenges and Solutions in Adopting VR in Mechanical Engineering

As technology continues to evolve at an unprecedented pace, the field of mechanical engineering is increasingly experiencing the transformative impacts of Virtual Reality (VR). Nevertheless, as is often the case with the application of novel technologies, there exist a number of challenges that must be overcome, each of which is accompanied by an emerging suite of solutions. This chapter delves into these obstacles and the creative strategies designed to mitigate them, in a quest to fully integrate VR into mechanical engineering.

10.1. Proficiency Gap in VR Technology

A significant challenge with adopting VR in mechanical engineering is the proficiency gap in the VR technology. This gap refers to the difference in technological skills and knowledge between the current generation of mechanical engineers and the newer generation who have grown up in an increasingly digitalized world.

To bridge this gap, a series of comprehensive VR trainings and workshops have been implemented, aiming to equip mechanical engineers with the necessary skills and knowledge to harness VR. These initiatives cover not only the basic mastery of VR principles and applications, but also delve into the teaching of programming languages and software used in creating and manipulating VR simulations.

10.2. Hardware and Software Compatibility

The diversity of hardware and software components that comprise VR systems can frequently lead to compatibility issues. It is not uncommon for certain VR software platforms to work optimally on specific hardware configurations only, thereby posing substantial challenges in setting up an inclusive and functional VR environment.

To surmount the compatibility issues, collaborations among VR technology developers, hardware manufacturers, and software developers are essential. Strides are being made to build universal software interfaces and adaptable hardware architectures, aiming to ensure seamless integration among all VR system components.

10.3. High Cost of VR Technology

One crucial barrier to the widespread adoption of VR in mechanical engineering is its high cost. The price of the hardware, specifically the head-mounted displays, data gloves, and motion tracking systems, coupled with the cost of maintaining the software, can be prohibitive for many organizations.

Nevertheless, solutions are also being devised to combat this impedance. Initiatives such as the development of cost-efficient VR equipment and the implementation of cloud-based VR systems promise to lower the barrier significantly. The latter, in particular, mitigates the need for high-end hardware, as heavy computations are offloaded to the cloud, potentially reducing the overall cost of VR systems.

10.4. VR Induced Cyber Sickness

Another issue is the cyber sickness caused by using VR technology.

Some users experience discomfort, dizziness, and even nausea when exposed to VR for prolonged periods, posing a significant challenge for widespread adoption.

This challenge is being addressed through the development of advanced VR hardware with improved motion tracking and display refresh rates. Additionally, software is being designed to minimize drastic viewpoint changes or sudden movements that can cause disorientation. Moreover, behavioral interventions, user education about the responsible use of VR, along with ample rest periods can also complement technology-based solutions.

10.5. Lack of Standardized VR Protocols in Mechanical Engineering

A lack of standardized protocols and guidelines for using VR in mechanical engineering serves as another barrier to adoption. Standards are essential to ensure proper and safe usage, adequate training, fair evaluation, and comparison of different VR applications.

To address this challenge, various regulatory bodies and institutions are working towards the development of standardized protocols. They aim to devise guidelines grounded in broad consensus among academics, practitioners, and stakeholders from different fields to provide a cogent framework for VR application in mechanical engineering.

Despite the challenges, the integration of VR in mechanical engineering holds substantial promise. These hindrances are fertile grounds for innovation, pushing boundaries to shape the future of mechanical engineering in an increasingly digital world. By addressing these challenges, we can steer the adoption of VR in mechanical engineering towards a more inclusive, efficient, immersive, and ingenious direction.

Chapter 11. Conclusion: Maximizing the Potential of VR in Mechanical Engineering

As this exploratory analysis of VR and its influence within mechanical engineering draws to a close, it is evident that technology is interweaving systemic advances into this dynamic, traditionally 'hands-on' discipline. What remains to be delineated, however, is understanding how to leverage this technological integration to its maximum potential in order to drive new advances and remain competitive in an increasingly digitized global sphere.

11.1. Approaching VR through a Systemic Lens

Integrating VR into mechanical engineering practices involves looking at it through a systemic lens. As a good starting point, it would require an in-depth understanding of VR's capabilities, limitations, possible applications and its broader impact. Simultaneously, it is vital to understand mechanics, mechanisms, material dynamics and more. Therefore, providing training and developing awareness programs could be an ideal way to begin this endeavor. Such a systemic perspective would, in turn, lead to more informed decisions, ultimately maximizing the effective and efficient integration of VR.

11.2. Effective and Efficient Training Programs

Promoting efficient integration of VR into mechanical engineering largely occurs through well-crafted interactive content. This process should enable engineers to get hands-on experience in a simulated environment. For instance, in the training module, engineers could be given tasks such as assembling and disassembling a mechanism in virtual spaces. The advantage of such training activities resides in the ability to risk-proof learning, allowing mistakes and detection of design flaws before actual manufacturing - thus saving resources and protecting the safety of personnel.

11.3. Leveraging VR for Design Optimization

VR has a profound implication for design optimization in mechanical engineering. By creating fully immersive simulations, designers can interact with their models in real time, fine-tuning designs based on the feedback from these interactions. Through VR, design optimization becomes an iterative process, enhancing the product's ergonomic and functional aspects while minimizing design revisions in the later stages of product development. Thus, VR can help in reducing both design time and cost.

11.4. Geographically Distributed Collaboration

In today's globalized world, engineering teams often collaborate from multiple geographical locations, creating a need for effective remote collaboration tools. VR can play an instrumental role in overcoming geographical barriers. Use of shared virtual spaces powered by VR can uphold real-time collaborations, enabling teams to work together on prototypes irrespective of their geographic locations. This not only reduces the time and resources spent on travel but also expedites product development and innovation.

11.5. Making the Most out of VR Investment

VR is a significant investment and needs a strategic roadmap to ensure that the organization attains the desired benefits from its deployment. Initially, organizations might explore VR applications in low stakes environments, such as training or communication, and once efficacy is ascertained, they may proceed towards incorporating it in high-stakes tasks like complex system design and virtual testing. Furthermore, companies can advance their utilization of VR by integrating it with other emerging technologies such as augmented reality, artificial intelligence and machine learning.

11.6. Advancing Safety and Efficiency

VR holds potential to improve safety in mechanical engineering. It can simulate hazardous situations and intricate processes, thereby providing a safe, controlled environment for training and hypothesis testing. In terms of efficiency, VR enables accurate and advanced visualization of complex systems and processes. This could lead to more realistic system analyses and predictions, thereby fostering a deeper comprehension and identification of possible system failures or inefficiencies prior to actual testing or implementation.

In conclusion, the potential of VR in mechanical engineering is vast and transformative. To exploit this to the fullest, stakeholders need to develop a comprehensive understanding of VR, initiate a systemic approach to integration, promote effective training, and instigate iterative design processes. From fostering remote collaboration to maximizing the return on VR investment, there is a multitude of ways in which VR can reshape mechanical engineering. The implementation and management of VR, however, must be done skillfully, with a keen eye on ensuring safety and efficiency. As we navigate through this digitized epoch, embracing the power of VR becomes not just a choice, but a compelling necessity for perpetuating progress and remaining relevant in the evolving landscape of mechanical engineering.

www.ingramcontent.com/pod-product-compliance
Lightning Source LLC
Chambersburg PA
CBHW072221290526
45794CB00007B/2830